JERUSALEM ARTICHOKE
Production and Marketing

Roby Jose Ciju

Contents

Jerusalem Artichokes or Sunchokes 4

1. Introduction 4

    1.1 Taxonomy 4

2 Origin and Distribution 4

3 Cultivars 5

4 Botanical Description 6

4.1 Plant 7

4.2 Flower 7

4.3 Tuber 7

5. Ecology 7

6. Nutritional Value 8

7. Production Requirements 9

    7.1   Climatic requirements for Jerusalem artichoke 10
    7.2   Soil requirements 10
    7.3   Propagation of Jerusalem artichoke 10
    7.4   Planting 10
    7.5   Fertilization 11
    7.6   Irrigation 11
    7.7   Weed Control 12
    7.8   Disease Management 12
    7.9   Insect Pest Management 12
    7.10   Interculture and Aftercare 13

8. Harvesting 13

9. Yield 14

10. Post Harvest Management 14

    10.1   Quality Indices 14
    10.2   Handling of Produce 15
    10.3   Storage 15
    Source: (Stanley J. Kays, 2008) 16
    10.4   Controlled Atmosphere (CA) Storage 16
    10.5   Retail Outlet Display 17
    10.6   Chilling Injury of Jerusalem Tubers 17
    10.7   Ethylene Sensitivity 17
    10.8   Respiration Rates 17
    10.9   Physiological Disorders during Storage 17
    10.10 Postharvest Disorders 18

11. Marketing Considerations 19

11.1	Grading	19
11.2	Packaging	19
11.3	Marketing	19

12. Uses of Jerusalem Artichoke 21

12.1	Food purposes	21
12.2	Sugar production	22
12.3	Alcohol production	22
12.4	Medicinal properties	22
12.5	Forage Production	23

13. Economics of Production 23

Bibliography *25*

List of Tables

Table 1: Taxonomy of Jerusalem artichoke 4

Table 2: Commercial Cultivars of Jerusalem artichoke 5

Table 3: Botanical Description of Jerusalem artichoke 7

Table 4: Nutritional Value of Jerusalem artichoke Tubers 8

Table 5: Irrigation Schedule 11

Table 6: Harvesting Parameters for Jerusalem Artichoke 13

Table 7: Quality Indices for Jerusalem artichoke tubers 14

Table 8: Optimum Storage Conditions for Jerusalem Artichokes 15

Table 9: Respiration Rates of Jerusalem artichoke Tubers at Different Temperatures 17

Table 10: Physiological Disorders of Jerusalem artichoke Tubers 18

Table 11: Postharvest Disorders of Jerusalem artichoke Tubers 18

Table 12: Marketing of Jerusalem artichoke Tubers 20

Table 13: Jerusalem Artichoke: Economics Of Production Per Acre 24

Jerusalem Artichokes or Sunchokes

1. Introduction

Scientific name of Jerusalem artichoke is *Helianthus tuberosus*. It belongs to the family Asteraceae (Compositae). Even though Jerusalem artichoke is a perennial plant, it is grown as an annual under commercial production practices. It is mainly grown for its edible root tubers which is a popular root vegetable in many parts of the world. Jerusalem artichokes are also known as sunchokes; sunroots; earth apples; and girasol. Jerusalem artichokes are very hardy plants with a mature height of 2.5 to 3 meters and a mature spread of 60 centimeters.

1.1 Taxonomy

Jerusalem artichoke is a member of Asteraceae family that includes plants like sunflower, daisy, and chrysanthemum. Botanical name of Jerusalem artichoke is *Helianthus tuberosus*. A detailed taxonomic classification of Jerusalem artichoke is as given in Table 1

Table 1: Taxonomy of Jerusalem artichoke

Kingdom: Plantae
Order: Asterales
Family: Asteraceae
Tribe: Heliantheae
Genus: Helianthus
Species: tuberosus

Source: USDA plants database

2 Origin and Distribution

Jerusalem artichoke is a native of North America, particularly the United States of America. Scientific researches on the origin of Jerusalem artichoke point out to the fact that Native American Indians have long been using Jerusalem artichoke as a major food. Over the years, Jerusalem artichoke has been introduced in both northern and southern hemispheres and gradually became naturalized in all temperate regions.

3 Cultivars

'Jerusalem White', Veitch's Improved Long White', 'Sutton's New White', 'Mammoth French White', 'French White Improved', 'Columbia' and 'LSD' are a few commercial cultivars those propagate vegetatively. Other commercial varieties are as shown in Table 2:

Table 2: Commercial Cultivars of Jerusalem artichoke

Stampede
Stampede is an early yielding short season variety of Jerusalem artichoke. The crop matures within 90 days with its flowering starts in July and tubers are ready in September. Stampede produces white colored tubers, each tuber weighing up to 1/2 lb each. Each plant may yield up to 10 lbs or more of tubers. Stampede is an extremely productive, high yielding variety of Jerusalem artichoke. Tubers are normally produced in big clusters near the main stem and hence they are easy to find while harvesting.
White Fuseau
White Fuseau is another highly productive variety of Jerusalem artichoke. But it is a late season variety. White Fuseau, as the name suggests produces white colored underground tubers which are large, long, crispy and knob-free with a thinner skin than other varieties. White Fuseau has one of the most favored flavors among the cultivated Jerusalem artichoke varieties and hence a much preferred selection for commercial cultivation. This variety is highly vigorous in growth habit and hence

spreads quickly.

Red Fuseau

Red Fuseau is another cultivated variety of Jerusalem artichoke which is preferred for its crispy, good flavored and thin, smooth skinned tubers. Red Fuseau produces red colored, small round and oblong shaped underground tubers in dense clusters near the main stem. It is an early yielding, highly productive variety of Jerusalem artichoke. It is also vigorous in growth habit with a height of up 10 ft. on maturity.

Clearwater: Clearwater is another high yielding variety of Jerusalem artichoke. Clearwater produces long thin tubers like those of white Fuseau tubers except that tubers are with few knobs. Clearwater has the most non-artichoke flavor compared to other commercial cultivars.

Seed Grown Jerusalem Artichoke: Jerusalem artichokes are sterile and rarely produce any seeds. Seedlings that are grown from seeds are very rare.

Red Rover

Red Rover is a prolific variety of Jerusalem artichoke and it grows up to 12 ft. in height and spreads quickly. Red Rover produces red colored smooth thin skinned tubers which are generally knob-free. Red Rover tubers are large with an average 1" in diameter and up to 6" in length. It ripens in October.

Waldspinel : Waldspinel Jerusalem artichoke produces small, red colored underground tubers with a bumpy surface

4 Botanical Description

Jerusalem artichoke is a perennial dicotyledonous plant and a detailed description of the plant is as given Table 3.

Table 3: Botanical Description of Jerusalem artichoke

4.1 Plant

Jerusalem artichoke is a tall, perennial herbaceous plant with erect, hairy stems and hairy, rough leaves. Though it is a perennial plant, it is often cultivated as an annual in commercial production. It grows up to a height of two to three meters. Leaves are with winged petioles; serrate-dentate; ovate to oblong shaped and are arranged in opposite directions on the lower stems; while alternate arrangement of leaves is seen towards the top of the plant.

4.2 Flower

Yellow colored flowers are produced terminal on the branches and resemble a sunflower in shape and size. Size of the flower head varies from 5 cm to 7.5 cm across that contains a yellow colored disk and ray florets up to 12 to 20. Flowering time is normally in July to August.

4.3 Tuber

Edible portion of a Jerusalem artichoke plant is its modified underground tubers which are botanically termed as 'rhizomes'. Tuber color varies from white to brown to red depending on the cultivar. Elongated and irregular shaped tubers are produced at the ends of the underground stems. Knobby or knob-free tubers are produced depending upon the variety. Size of each tuber varies from 7.5 cm to10 cm in length and 3 cm to 5 cm in thickness.

5. Ecology

Jerusalem artichoke is a hardy plant that can be grown in any soils and under any climatic conditions though temperate climate is generally preferred. The plant is susceptible to frost injury and the first frost normally kills the vegetative growth; however tubers withstand frost injury. It grows best in full sun and with plenty of water. Plants are sensitive to day length and hence they do not flower in northern Europe where days are very short. Plants require longer days from planting to maturation period and short days during tuber formation.

6. Nutritional Value

Jerusalem artichokes are high in potassium, iron, dietary fibers, niacin, thiamine, copper and phosphorus. An illustration of nutritional value per 150 grams of serving is given in Table 4.

Table 4: Nutritional Value of Jerusalem artichoke Tubers

(In 150 grams of serving)

Calories	109 (456 kJ)
Total Carbohydrate	26.2 g
Dietary Fiber	2.4 g
Sugars	14.4 g
Total Fat	0.0 g
Total Omega-6 fatty acids	1.5 mg
Protein	3.0 g
Vitamin A	30.0 IU
Vitamin C	6.0 mg
Vitamin E (Alpha Tocophcrol)	0.3 mg
Vitamin K	0.2 mcg
Thiamin	0.3 mg
Riboflavin	0.1 mg
Niacin	2.0 mg

Vitamin B6	0.1 mg
Folate	19.5 mcg
Pantothenic Acid	0.6 mg
Choline	45.0 mg
Calcium	21.0 mg
Iron	5.1 mg
Magnesium	25.5 mg
Phosphorus	117 mg
Potassium	643 mg
Sodium	6.0 mg
Zinc	0.2 mg
Copper	0.2 mg
Manganese	0.1 mg
Selenium	1.1 mcg
Cholesterol	0.0 mg
Water	117 g
Ash	3.8 g

Source: USAD Nutrient database

7. Production Requirements

Jerusalem artichoke plants are very easy to grow as they are very hardy plants and once roots are established on soil, they tend to grow vigorously with little nutrition and less care. A thorough soil preparation is normally done in commercial cultivation of Jerusalem artichokes. Soil is prepared well by adding and mixing bulk quantities of compost and well rotted farm yard manure liberally into the top soil and then adding a little lime just before planting Jerusalem artichoke tubers. After planting is done, soil is kept free from weeds all throughout its production duration. It is necessary to keep the soil slightly moist always.

7.1 Climatic requirements for Jerusalem artichoke

Jerusalem artichokes are better adapted to cool climates. Temperature requirements vary from 65 to 80°F and rainfall requirements vary from 50 inches or less. Crops are grown both as rainfed and irrigated crops. Irrigation may be necessary for rainfed crops also, if soil is dry.

7.2 Soil requirements

Jerusalem artichoke is adapted to various soil types but the best soil is fertile sandy loams or well-drained, slightly alkaline soils. Generally speaking, soils suitable for potato and corn production are suitable for Jerusalem artichoke production also. Water logging must be avoided but soil moisture must be well above 30% of field capacity during the tuber formation period which starts from late August to early September and lasts up to November.

7.3 Propagation of Jerusalem artichoke

Vegetative propagation via seed tubers is generally preferred in Jerusalem artichoke. Tubers or pieces of tubers containing two or three vigorous buds and weighing approximately 50-60 grams are used for the propagation of Jerusalem artichoke. Tubers start sprouting after two to three weeks of planting. During plant establishment, grass and weed problems will be reduced by shading since plants grow over 6 feet high. Tubers begin to form in August and may become 4 inches long and 2 to 3 inches in diameter upon full growth.

7.4 Planting

Recommended planting rate is approximately 1.5 tons of seed tubers per hectare which yield up to 25,000 to 30,000 plants per hectare. Ideal planting time is spring through early summer (January to March). Ideal spacing is 12 to 25 inches between plants and 30 to 35 inches between rows at a planting depth of 2 to 4 inches. Tubers begin to form in August.

7.5 Fertilization

Fertilizer application helps produce a better yield. At the time of land preparation an application of FYM (farm yard manure) or compost @25 tons /hectare improves top soil fertility. For better results, apply adequate quantities of neem cake along with FYM. As a rule, fertilizer requirements of Jerusalem artichoke plants are same as that of potatoes. Generally it is suggested that 1000 Kg per hectare of 6-12-6 NPK should be broadcasted in the row. This rate may be increased on low fertile soils.

7.6 Irrigation

Sunny position and water are two essential requirements for the successful production of Jerusalem artichokes. Jerusalem artichokes need to be watered deeply to force the production of large tubers. Less watering tends to produce less and smaller tubers. An illustration of irrigational schedule is as given in Table 5.

Table 5: Irrigation Schedule

First irrigation	Immediately after planting a light irrigation is given
Subsequent irrigations	As and when needed, depending on soil and weather conditions
Stop Irrigation	At the time of maturity of tubers, irrigation is detrimental

7.7 Weed Control

Mechanical weed control is generally recommended for Jerusalem artichokes. Manual removal of weeding, by carefully pulling of any weeds that appear, until the plants get well established is highly recommended.

7.8 Disease Management

Sclerotinia rot is a major problem in Jerusalem artichoke. Other fungal diseases that affect Jerusalem artichoke include downy mildew, rust and southern stem blight.

7.9 Insect Pest Management

Puccinia helianthi is the most serious pest that attack Jerusalem artichoke plants.

Control of Pests and Diseases

The best control measure to keep diseases and pests away is to keep a healthy ecosystem in the growing fields. Key components of a healthy agroecosystem include:

- Proper land preparation prior to planting
- Improving soil fertility by the use of organic manures and fertilizers, organic mulches, and rotation with cover crops
- Selection of varieties adapted to the location
- Proper irrigational management to prevent Drought or Waterlogging
- Intercropping with compatible crops
- Understanding about pest life-cycles

In cases of severe insect-pest infestations, burning diseased plants and a change of locality is recommended for the absolute control of the pest. In cases of severe diseases, first option is to use organic control measures since there are no major registered chemicals recommended for Jerusalem Artichokes.

7.10 Interculture and Aftercare

Regular hoeing and weeding is recommended until tuber formation and thereafter hoeing and weeding must be stopped. 2–3 hoeing and weeding are recommended. Generally manual weeding is recommended. Mulching with black polythene also controls weed growth.

Since Jerusalem artichoke plants remain dormant during winter season, cut the flower stalks off at the ground level in order to help this vegetable survive the winter. Then cover the plants with proper mulching to protect them from frost injury. Mulch can be removed during spring season.

8. Harvesting

Harvesting of Jerusalem artichokes is similar to that of potatoes. Only major consideration is that harvest the product at its correct maturity stage. A detailed illustration of harvesting process for Jerusalem artichokes is given in Table 6.

Table 6: Harvesting Parameters for Jerusalem Artichoke

Harvesting Method	Manual harvesting using a potato digger
Field Packing	In gunny bags or jute bags
Harvesting Stage	• When the tops of the plants wilt or die
	• Normally during early winter (November to January)
How to Harvest	• Tops should be cut down to 12" above the ground with a mower
	• Then plough open the furrow
	• Pick up the tubers using a potato digger

	• Place the harvested tubers in field containers or jute bags
	• Transport the freshly harvested produce to the pack houses immediately
	• Harvest the tubers in 4 or 5 months as it is best to leave them in the soil and harvest as and when needed
Other Considerations	• Harvesting should be done in early hours of morning to avoid extreme heat building in tubers.

9. Yield

Average tuber yield ranges between 5 and 10 tons per acre. Alcohol yield is at 60–100 liters/metric tons of tubers. Yield of tops for forage is between 10 and 15 tons per acre.

10. Post Harvest Management

10.1 Quality Indices

Tuber size and tuber shape are major quality parameters to be considered while marketing the product for target markets. An illustration of quality parameters is given in Table 7.

Table 7: Quality Indices for Jerusalem artichoke tubers

Size, shape and color	Uniform shape, size and color typical of variety
Texture of Tubers	Firm
Appearance	1. Skin of tubers should be free of bruises, injuries and similar damages 2. Tuber should be clean; free of soil and dirt 3. Freedom from defects such as cracks, decay, sunburn

10.2 Handling of Produce

Since the skin of Jerusalem artichokes is very thin, care should be taken during post harvest handling of the tubers to avoid skinning, cuts and bruises. The skin is also susceptible to rapid moisture loss; hence it is advised to store the tubers under favorable low temperature soon after harvest. Precooling is not required for Jerusalem artichoke tubers.

10.3 Storage

There are mainly three storage options recommended for storing Jerusalem artichoke tubers. An illustration of the same is in Table 8.

Table 8: Optimum Storage Conditions for Jerusalem Artichokes

In situ field storage	1. In field storage tubers are left in the ground and harvested as and when required 2. Field storage is mainly dependent upon locational factors 3. Recommended for northern hemisphere production areas where low soil temperatures prevail throughout the winter

	4. Recommended for locations where soil is sandy, and well-drained
Cold storage or Refrigerated storage	1. Highly effective, but an expensive option
2. Recommended for seed tubers and tubers for fresh markets
3. Ideal cold storage temperature is 32 to 34^0 F ($0°C$ to $2°C$)
4. Ideal humidity is 90 to 95% relative humidity (RH)
5. At low RH, tubers shrivel and decay begins
6. Storage life is up to 6 months
Common storage in root cellars	1. Cooling of tubers is obtained from the natural outdoor air and/or soil kept at low temperatures
2. Recommended when the tubers must be harvested during fall season, i.e. before winter begins
3. Also recommended when cold storage/ refrigeration is not available or very expensive

Source: **(Stanley J. Kays, 2008)**

10.4 Controlled Atmosphere (CA) Storage

Several researches on the effect of controlled atmosphere storage on the quality of Jerusalem tubers have revealed that CA storage of Jerusalem tubers at 22.5% CO_2 + 20% O_2 decreases inulin degradation process in the tubers, thus adds some storage value.

10.5 Retail Outlet Display

For retail marketing, Jerusalem artichoke tubers may be displayed in refrigerated product display packages. Loose tubers may be displayed under high Relative Humidity conditions facilitated by misting process.

10.6 Chilling Injury of Jerusalem Tubers

Jerusalem tubers are not sensitive to chilling. However, freezing at -5 °C (23 °F) causes little damage though not lethal but freezing at -10 °C (14 °F), may result in rapid deterioration of the tubers.

10.7 Ethylene Sensitivity

Jerusalem tubers are not sensitive to ethylene.

10.8 Respiration Rates

A detailed illustration of respiration rates of tubers at different temperatures is given in Table 9.

Table 9: Respiration Rates of Jerusalem artichoke Tubers at Different Temperatures

Temperature	mg CO_2 kg^{-1} h^{-1}	Rate of dry mater loss (g kg^{-1} day^{-1})
0 °C	10.2	0.162
5 °C	12.3	0.201
10 °C	19.4	0.317
20 °C	49.5	0.801

To get mL kg^{-1} h^{-1}, divide the mg kg^{-1} h^{-1} rate by 2.0 at 0 °C (32 °F), 1.9 at 10 °C (50 °F), and 1.8 at 20 °C (68 °F). To calculate heat production, multiply mg kg^{-1} h^{-1} by 220 to get BTU per ton per day or by 61 to get kcal per metric ton per day

10.9 Physiological Disorders during Storage

Major physiological disorders of Jerusalem artichoke tubers under storage are desiccation of tubers, rotting and sprouting of tubers. An illustration of major physiological disorders of Jerusalem tubers under different storage conditions is given in Table 10.

Table 10: Physiological Disorders of Jerusalem artichoke Tubers

Desiccation	1. A significant storage problem
	2. Storage at high RH prevents tubers from rapid desiccation
Rotting	1. Major cause is excess humidity
	2. Maintain optimum RH for controlling rotting of tubers in the storage
Sprouting	1. Sprouting of tubers in the storage is mainly due to the presence of excess moisture
	2. Control measure: Maintain optimum RH

10.10 Postharvest Disorders

Major pathological disorders are botrytis or grey mold decay and Sclerotinia Rot. An illustration of major physiological disorders of Jerusalem tubers under different storage conditions is as shown in Table 11

Table 11: Postharvest Disorders of Jerusalem artichoke Tubers

Botrytis Rot	1. Caused by Botrytis cinerea i.e. Botrytis decay is a fungal infection
	2. Control measures
	a. Prevention of bruises/wounds on the tuber surface help reduce its incidence .i.e. Minimize mechanical damage
	b. Field sanitation reduces infection
	c. Store tubers at 0 to 2 °C (32 to 34 °F) for preventing

	botrytis infection
	d. Remove all diseased tubers from the storage to check further infection
	e. Proper RH control in storage also yields better results
Sclerotinia Rot	1. Caused by Sclerotinia sclerotiorum
	2. Control measures: Same as above

(http://www.ba.ars.usda.gov/hb66/077jerusalem.pdf, 2012)

11. Marketing Considerations

11.1 Grading

As of now, there are no national or international standards for grading Jerusalem artichoke tubers. As a rule, larger tubers with smooth surfaces are considered as the best grade.

11.2 Packaging

For bulk marketing, gunny bags or jute bags are generally used for packing Jerusalem artichoke tubers. Perforated polyethylene bags are used for packaging the tubers for retail marketing. For long term storage in cold storages, tubers are peeled and sliced before placing in cold water with lemon juice for preventing discoloration. Thereafter sliced tubers are blanched in boiling water for few minutes and then cooled and dried before freezing them for half an hour. Frozen tubers are then packed in air tight bags before sealing and labeling them. These frozen Jerusalem artichoke tubers can be stored up to 6 months.

11.3 Marketing

Tubers are marketed as fresh produce as well as processed products. Fresh Jerusalem artichoke tubers are generally available in the market from November to

April. Processed and frozen tubers are available at any time. An illustration of major marketing options is given Table 12.

Table 12: Marketing of Jerusalem artichoke Tubers

Local markets	There is a small market for Jerusalem artichokes in India and other South Asian countries. However the product has a great potential for local marketing with the flourishing hotel industry. However it is advised that growers should identify a buyer or buyers before starting large scale production Jerusalem tubers.
Export markets	Asia, Mexico, Europe and Central America

Specialty markets	Since Jerusalem artichoke tubers are considered as an exotic/specialty vegetable in many parts of the world, these tubers can be marketed for specialty markets. Organically produced tubers may fetch a premium and may have more demand among health-conscious buyers.
Marketing as a Branded Product	Since Jerusalem artichokes are a rare vegetable, commercial producers may brand their product and sell them in niche markets.
Internet sales	With the growth rate that observed in the field of e-commerce in recent times, producers may explore the options for selling their products through internet.

(Parker, 2009)

Quarantine Issues

There is no quarantine issues associated with the marketing of Jerusalem tubers. Properly graded and packed tubers may be marketed in the local markets or export markets with or without a single fumigation treatment with recommended fumigants.

12. Uses of Jerusalem Artichoke

12.1 Food purposes

Jerusalem artichoke is mainly cultivated for its underground tubers which are edible and are used as root vegetables. Raw tubers resembles potatoes in consistency and texture but more sweet and nuttier in flavor. Tubers have a crispness that resembles water chestnuts and therefore most suitable for salad preparation. Jerusalem artichokes are used for pickling purposes and can also be cooked like potatoes. Thin slices of tubers are used for preparing chips. The foliage of Jerusalem artichoke plants makes a good forage crop for the livestock. Stems and leaves are rich in fats, protein and pectin, and make good forage and silage.

12.2 Sugar production

Jerusalem artichokes store carbohydrates in the form of inulin and are an important source of fructose. Since tubers are rich in fructose, tubers may be used for sugar extraction though these tubers are not yet commercially exploited for sugar production.

12.3 Alcohol production

Jerusalem artichoke tubers are widely used in France, Germany and many other European countries for producing alcoholic beverages including beer and wine. Jerusalem artichoke has a high content of easily hydrolyzed inulin and hence may be used for ethanol production. But a detailed research is needed for formulating an economically viable method for ethanol extraction from Jerusalem artichoke tubers.

12.4 Medicinal properties

Though most of the root tubers store carbohydrates as starch, Jerusalem artichokes and other artichokes store carbohydrates as inulin. Therefore Jerusalem artichoke tubers serve as the best substitutes for starches in many diet preparations. Besides, inulin is easily digestible and helps body in increasing calcium absorption. Another important point is, consumption of Jerusalem tubers does not increase blood sugar and therefore highly recommended for the diabetic patients as a part of their regular diet and hence it is also known as '*diabetics potato*'. Jerusalem artichoke can also be used an aphrodisiac, diuretic, stomachic, and tonic, Jerusalem artichoke is also used as a remedy for rheumatism (Loes, 2000).

12.5 Forage Production

Tops of Jerusalem artichokes are used for feeding livestock. Leaves are rich in TDN (total digestible nutrients); however forage quality is comparatively low than that of alfalfa.

13. Economics of Production

Cost of production of Jerusalem artichoke tubers varies depending upon locality, soil type, and other relevant parameters. However it is estimated that on an average, USD (US Dollars) 1400 to 2000 is spent toward the production of approximately 10 tons of tubers from an area of one acre.

Major production expenses include cost of seed tubers; labor cost involved in field preparation and planting process, and costs of cultural management; cost of fertilizers and pesticides if required, labor cost involved in irrigation and disease-pest management and weed management; costs of harvesting process and postharvest management. As a general rule labor cost may amount up to 50 % of total production costs. Major income obtained is through the sales of seed tubers and fresh tubers.

Major Risks Involved in the Production of Jerusalem artichoke Tubers

Three major risks faced in all cases of agricultural production are Production-related risks; Price-related risks and Finance-related risks. Risks associated with the production are unexpected natural calamities, insect-pest damage and disease incidences. Price risks include unexpected fall in product prices while major finance-related risks are sudden increase in loan interest rates. Growers are advised to be well-prepared to manage these risks before embarking on a large-scale commercial production by gathering as much information as possible on production practices, price trend of the product over a period of time and other relevant market information.

An estimated economics for the production of Jerusalem artichoke crop per acre area is as shown in Table 13.

Table 13: Jerusalem Artichoke: Economics Of Production Per Acre

Parameters	Estimated cost**
Cost of seeds/tubers	$ 1,000/acre
Cutting and planting	$60 to 150/acre
Cost of cultivation	$25 to 50/acre
Cost of harvesting	$250 to 400/acre
*Miscellaneous expenses	$100 to 300

**all figures are in dollars

*miscellaneous expenses include storage, transportation, and supplemental seed stock

Bibliography

Loes, M. W. (2000). *Healing Power of Jerusalem Artichoke Fiber* . USA: Freedom Press.

Stanley J. Kays, S. N. (2008). *Biology and chemistry of Jeruslaem artichoke:helianthus tuberosus L.* USA: CRC Press.

Parker, P. M. (2009). *The 2011 World Forecasts of Fresh or Dried Arrowroot, Salep, Jerusalem Artichokes, Sweet Potatoes, and Other Roots and Tubers with High Starch or Inulin Export Supplies.* USA: ICON Group International Inc.

http://www.ba.ars.usda.gov/hb66/077jerusalem.pdf. (2012, January Tuesday). Retrieved January Tuesday , 2012, from USDA: http://www.ba.ars.usda.gov/hb66/contents.html

Made in the USA
Lexington, KY
07 June 2012